Advanced Cardiovascular Life Support (ACLS)

A Comprehensive Guide Covering the Latest Guidelines

Workbook

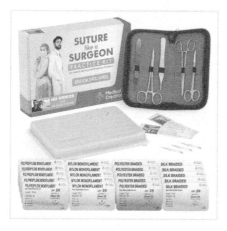

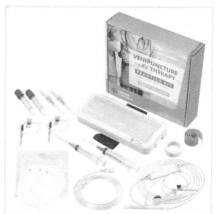

Contents

FREE GIFT

GET ONE OF THESE EBOOKS FOR FREE:

Medical Reference Pamphlet
ACLS ebook
Medical Terminology Digital Pamphlet
Neurology ebook
Mini Medical Dictionary
ECG - Digital Reference Pamphlet

Scan the following QR Code:

You will be redirected to our website.
Follow the instructions to claim your free gift.

INTRODUCTION

The principles and practice of advanced cardiovascular life support (ACLS) are at the forefront of the battle to save precious lives every day. Understanding these principles and adopting them into practice gives healthcare providers a unique set of life-saving tools that are steeped in evidence-based medicine.

Healthcare professionals can learn all about the latest information on ACLS in our very well-received ACLS textbook [**can insert a link here to the ACLS guide**]. The textbook is succinct yet covers all important areas and algorithms of ACLS. It can be used as your essential study guide toward earning ACLS certification.

This workbook complements the ACLS textbook. If you're feeling confident about your knowledge of ACLS principles, you can start with this workbook. Alternatively, if you've just finished our ACLS textbook and are looking for practice questions and exercises to test your knowledge, then this workbook is exactly what you need.

Good luck with the exercises!

UNIT I

The ACLS Chain of Survival

EXERCISE 1 – FILL IN THE BLANKS

1. The chain of _____ is a series of actions that can decrease the mortality rate following cardiac arrest.

2. It is well recognized that early initiation can increase survival after a cardiac emergency. For early initiation of life support measures, two key initiatives are required: _____ and alerting for necessary resources.

3. In the out-of-hospital chain, the first thing to do is to _____ emergency services.

4. The goal of the rapid response team is to _____ the occurrence of respiratory or cardiac arrest.

5. A sudden deterioration in mental status in a patient is a criterion to alert the _____ team.

6. If respiratory or cardiac arrest has already occurred, the _____ team is called to resuscitate the patient.

7. The 2020 AHA guidelines have placed greater emphasis on encouraging layperson-initiated _____.

8. Delivering electric current using a _____ can 'shock' the heart into regaining its normal rhythm.

9. The type of defibrillator most commonly used in out-of-hospital settings is called _____.

10. In a hospital, a _____ defibrillator is commonly used.

11. Once the patient's life is saved, post-cardiac arrest care aims to _____ normal body function.

12. The last link in the chain, _____, signifies the importance of continued support to patients and their families.

EXERCISE 2 - MCQs

13. Like any chain, the chain of survival is only as strong as its:

    a. execution sequence
    b. weakest link
    c. final outcome
    d. strongest link

14. In the out-of-hospital chain, a layperson may recognize that an emergency exists only when the patient:

    a. calls for help
    b. moves
    c. collapses
    d. shouts

15. In the in-hospital chain of survival, alerting the emergency team is which step?

    a. first
    b. second
    c. third
    d. fourth

16. In a hospitalized 55-year-old man with a history of heart disease, you find that the heart rate is 145 bpm and respiratory rate is 30 breaths per minute. Which medical team would you immediately notify:

    a. code blue team
    b. code red team
    c. rapid response team
    d. anesthesia team

17. The code blue team will be called in all of the following situations except:

    a. Patient is unresponsive
    b. Patient stops breathing
    c. Patient's blood pressure drops
    d. Patient has no pulse

18. Despite the established efficacy of early cardiopulmonary resuscitation, the AHA estimates that the percentage of collapsed patients who receive layperson-initiated CPR is:

    a. less than 10%
    b. less than 20%
    c. less than 40%
    d. more than 60%

19. Irrespective of the number of rescuers, the ratio of compressions to rescue breaths for adults is:

    a. 15:2
    b. 30:2
    c. 45:2
    d. 60:2

5

20. In children, if a single rescuer is present, the ratio of compressions to rescue breaths is:

    a. 15:2
    b. 30:2
    c. 45:2
    d. 60:2

21. In children, if two rescuers are present, the ratio of compressions to rescue breaths is:

    a. 15:2
    b. 30:2
    c. 45:2
    d. 60:2

22. Defibrillation is effective when it is performed within how many minutes of an arrest?

    a. 3 to 5
    b. 6 to 8
    c. 8 to 10
    d. 10 to 15

23. For out-of-hospital arrests, the emergency response team must ideally consist of:

    a. two providers trained in ACLS and two in BLS
    b. two providers trained in ACLS and one in BLS
    c. one provider trained in ACLS and one in BLS
    d. one provider trained in ACLS and three in BLS

Anatomy and Physiology of the Heart and Normal Electrocardiogram

EXERCISE 1 - FILL IN THE BLANKS

1. Only _____ blood passes though the right heart.

2. The pulmonary veins open into the _____ atrium.

3. The activity that occurs in the heart during a single heartbeat is referred to as the _____.

4. During systole, the left ventricle pushes blood into the _____.

5. The _____ wave is a small wave which represents atrial depolarization.

6. The _____ wave represents ventricular repolarization.

7. To get an estimate of ventricular rate, divide 1500 by the number of small squares between identical points on two consecutive _____ waves.

8. The normal PR interval should be _____ seconds.

9. The QRS complex duration should be around _____ seconds.

10. If fibers outside the SA node stimulate the heart, it can result in _____ beats, which will appear as irregularities on the ECG.

EXERCISE 2 - MCQs

11. The sinoatrial (SA) node generates electrical impulses at an average rate of how many per minute?

   a. 20 to 40
   b. 40 to 60
   c. 60 to 100
   d. 100 to 150

12. If the distance between successive 'R' peaks varies, the rhythm is:

   a. undetectable
   b. regular
   c. irregular
   d. fast

13. What's the approximate heart rate (bpm) in the following ECG:

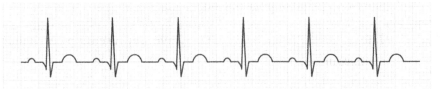

   a. 63
   b. 73
   c. 83
   d. 93

14. When checking consecutive P waves on an ECG reading, which of the following indicates a problem:

    a.   All cycles/beats have them
    b.   All have identical shape
    c.   Some are inverted
    d.   All are followed by a QRS complex

15. A prolonged PR interval most likely indicates blockage in transmission at the

    a.   SA node
    b.   AV node
    c.   Bundle of His
    d.   Bundle branch

16. If a QRS complex is prolonged, it most likely indicates blockage in transmission at the:

    a.   SA node
    b.   AV node
    c.   Bundle of His
    d.   Bundle branch

17. A raised ST segment on the ECG is a classic sign of:

    a.   myocardial ischemia
    b.   myocardial infarction
    c.   pulmonary embolism
    d.   hypokalemia

18. Peaked T waves are generally associated with:

    a.   hyperkalemia
    b.   hypocalcemia
    c.   hyperthermia
    d.   bundle branch block

EXERCISE 3 – LABEL THE PICTURE

The following image shows the main chambers of the heart and major components of the conducting system. Can you label them?

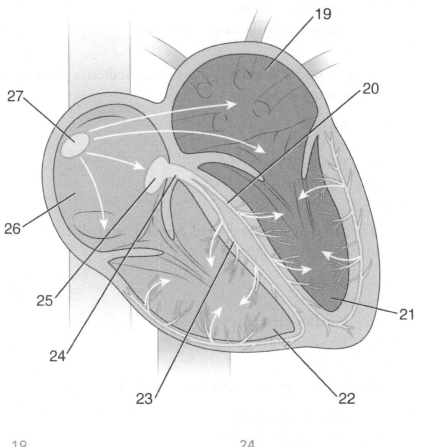

19. _____   24. _____

20. _____   25. _____

21. _____   26. _____

22. _____   27. _____

23. _____

The ACLS Survey

CHAPTER 1:

The Adult ACLS Algorithm

EXERCISE 1 – FILL IN THE BLANKS

Set A

The following table contains eight sets of options as life support measures.

Assess airway, perform advanced airway management	Rescue breathing	Check for breath sounds Deliver 100% oxygen Capnography to assess CPR quality	Basic airway management
Chest compressions	Assess heart rhythm, use defibrillator as indicated Assess and treat reversible causes	Gain parenteral access to deliver ACLS medication	Use of AED if available at the scene

Use the options provided above to fill out the blanks in the table below:

STEPS	INITIAL C-A-B-D STEPS (USUALLY UNDERTAKEN AS PART OF BLS)	ADVANCED C-A-B-D STEPS (ACLS COMPONENTS)
C – Circulation	1.	2.
A – Airway	3.	4.
B – Breathing	5.	6.
D – Defibrillation; Differential diagnosis	7.	8.

Set B

9. The AHA 2020 guidelines recommend that laypersons initiate early _____ in all presumed cases of cardiac arrest as the risk of harm to the patient is low if the person is not actually in cardiac arrest.

10. The current preference for a C-A-B approach over the A-B-C approach highlights the priority given to _____ over oxygenation.

EXERCISE 2 - ADULT ACLS ALGORITHM

Complete the labeling in the adult ACLS algorithm shown below.

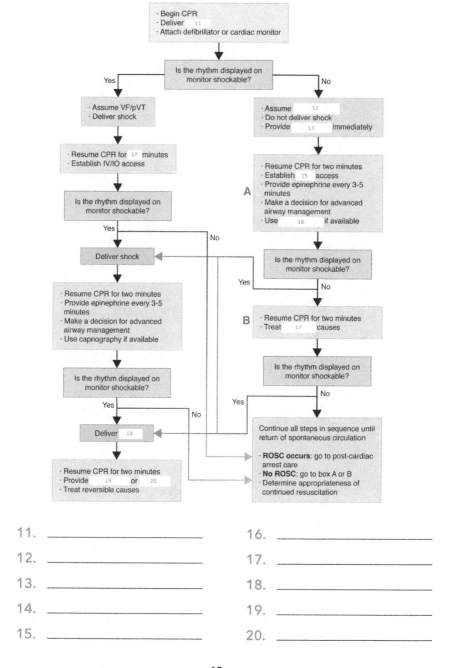

11. _____	16. _____
12. _____	17. _____
13. _____	18. _____
14. _____	19. _____
15. _____	20. _____

CHAPTER 2:

ACLS Survey 1 - Circulation

EXERCISE 1 - FILL IN THE BLANKS

Set A

1. The AHA recommends the _____ route for drug administration only when intravenous access is unsuccessful or not feasible.

2. Maintaining circulation is the first and most important priority of ACLS. It is primarily achieved by chest compressions and secondarily by delivering _____ drugs.

3. After each compression, allow complete chest _____.

4. Prior to placing an advanced airway, the ratio of compressions to mouth breaths is _____.

5. Unless absolutely necessary, chest compressions must not be _____, so as to avoid compromising perfusion.

6. The 2020 AHA guidelines suggest that it is reasonable to use real-time audiovisual feedback devices when available to optimize _____ quality.

7. In the active compression-decompression technique, a handheld suction device can potentially increase the negative _____ pressure during each chest recoil, which in turn can improve venous return and cardiac output.

Set B

8. To visualize the veins better when obtaining IV access, apply a tourniquet four to six inches _____ to the site of access.

9. The tourniquet must not remain for longer than _____ minutes.

10. As a general rule, when drugs are given intravenously, the entire drug must be delivered as a _____ and flushed with 20 ml of saline or any other IV fluid.

11. Epinephrine causes arteriolar constriction by stimulation of _____ receptors.

12. If a shockable rhythm is present, _____ must be administered after the second shock has been delivered.

13. The AHA does not recommend _____ as an alternative to epinephrine, as there is no benefit to either its substitution for or addition to epinephrine.

14. An alternative to amiodarone to manage VF/VT is _____.

EXERCISE 2 – MCQs

Set A

15. The depth of each compression must be at least:

    a. 1 inch (2.5 cm)
    b. 2 inches (5 cm)
    c. 3 inches (7.5 cm)
    d. 4 inches (10 cm)

16. The compressions must be performed at a rapid rate of approximately how many compressions per minute?

    a. 50 to 80
    b. 80 to 100
    c. 100 to 120
    d. 120 to 150

17. To avoid sub-optimal compressions due to fatigue, the rescuer performing compressions must switch over with the rescuer providing rescue breaths every:

    a. 1 minute
    b. 2 minutes
    c. 3 minutes
    d. 5 minutes

18. Quantitative waveform capnography is the preferred quantitative method to assess CPR quality and can be utilized if advanced airway management has been completed and an endotracheal tube is in place. During CPR, the rescuer must attempt to maintain ETCO2 levels between:

   a.   0 – 10 mmHg
   b.   10 – 20 mmHg
   c.   20 – 30 mmHg
   d.   35 – 45 mmHg

19. If continuous arterial blood pressure monitoring is available during CPR, the rescuer should attempt to maintain the relaxation pressure values above:

   a.   10 mmHg
   b.   20 mmHg
   c.   30 mmHg
   d.   40 mmHg

20. Manual compressions are subject to decrease in quality due to fatigue and human error. It was assumed that this could be overcome with mechanical devices. However, studies have not proven superiority of these devices over manual CPR. The AHA does not recommend the routine use of mechanical CPR devices. However, these devices may be considered if:

   a.   Delivering manual compressions is not possible
   b.   Delivering manual compressions is dangerous to the provider
   c.   The patient is highly infectious
   d.   All of the above.

21. Interposed abdominal compression CPR is a technique which can be carried out if at least how many trained rescuers are available?

    a. 1
    b. 2
    c. 3
    d. 4

Set B

22. When obtaining IV access, remove the cover from the IV needle. Using your dominant hand, hold it with its bevel facing upwards. What angle should the needle be making with the skin?

    a. 15°
    b. 30°
    c. 45°
    d. 60°

23. Intraosseous (IO) access may be the only option if the cardiac arrest is due to:

    a. Hypothermia
    b. Hyperkalemia
    c. Ventricular fibrillation
    d. Circulatory shock

24. The most convenient site for IO access is:

    a. anterior tibia
    b. anterior fibula
    c. anterior femur
    d. anterior patella

25. Which of the following is no longer a recommended route of drug administration in ACLS?

    a. IV
    b. IO
    c. Central line
    d. Intracardiac

26. The dosage of epinephrine to be given intravenously is 1 mg. What should be the dilution?

    a. 1:10
    b. 1:100
    c. 1:1000
    d. 1:10,000

27. The effect of amiodarone is that it:

    a. Constricts the proximal aorta
    b. Increases venous return
    c. Increases intrathoracic pressure
    d. Slows down heart rhythm

28. The initial bolus dose of amiodarone given intravenously is:

    a. 10 mg
    b. 50 mg
    c. 100 mg
    d. 300 mg

CHAPTER 3:

ACLS Survey 2 & 3 – Airway, Breathing and Ventilation

EXERCISE 1 - FILL IN THE BLANKS

1. If the neck is immobilized or neck injury is suspected, _____ maneuver is performed instead of the head-tilt, chin-lift maneuver.

2. Immediately after suctioning, ventilation must begin and the patient must be given _____ oxygen.

3. The oropharyngeal airway is only used when the patient is _____.

4. In a conscious or semi-conscious patient, use the _____ airway.

5. During bag-mask ventilation, attempt to deliver a volume of _____ with each breath.

6. Correct placement of an endotracheal tube can be confirmed by ventilating with an ambu-bag and auscultating for _____ sounds.

7. A _____ is an alternative to the ET tube and is preferable when a laryngoscope is not available or neck injury is suspected.

8. The Combitube is also known as the _____ tube.

9. After any type of advanced airway placement, use _____ to confirm that the airway is in place and to monitor CPR quality.

EXERCISE 2 - MCQs

10. In a patient requiring resuscitation, the airway may be obstructed by secretions such as saliva, blood or vomit. These can be cleared up using:

    a. jaw-thrust maneuver
    b. endotracheal intubation
    c. suctioning
    d. ventilation

11. The oropharyngeal airway is a stiff, J-shaped plastic device which can prevent:

    a. ventilation
    b. vomiting
    c. suctioning
    d. tongue fall back

12. The appropriate size of the oropharyngeal airway can be selected by placing the device at the side of the patient's face, parallel to the sagittal plane. The device should ideally extend from the corner of the mouth to the:

    a. earlobe
    b. neck
    c. angle of the mandible
    d. chin

13. You are attempting to place a nasopharyngeal airway in a patient. However, when you try to insert the tube through the nostril, you encounter resistance as if the nostril is blocked. What should be your next step?

    a. Apply more force until the tube passes through
    b. Try an oropharyngeal airway
    c. Deliver 100% oxygen
    d. Try the other nostril

14. Which of the following is a good sign that ventilation is adequate?

    a. cyanotic lips
    b. chest rise and fall
    c. rising blood pressure
    d. rapid pulse

15. Oxygen saturation must be monitored through a pulse oximeter, and the saturation must be maintained at or above:

    a. 84%
    b. 88%
    c. 94%
    d. 98%

16. The need for an advanced airway must be balanced against the risk of:

    a. infection
    b. capnography monitoring
    c. cyanosis
    d. CPR interruption

17. Which of the following is a device that is needed to insert an endotracheal tube?

    a. stethoscope
    b. laryngoscope
    c. suction tube
    d. oropharyngeal tube

18. In general, adult women will require an endotracheal tube of a size between:

    a. 6 and 7
    b. 7 and 7.5
    c. 7.5 and 8
    d. 8 and 9

19. The biggest drawback to endotracheal tube placement is that it:

    a. is an irreversible procedure
    b. takes too long to place
    c. requires appropriate training and experience
    d. is incompatible with capnography

20. As a general guide, which size of a laryngeal mask airway fits most adult men?

    a. 4
    b. 5
    c. 6
    d. 7

21. The Combitube is not generally preferred over other airway options, as it:

    a. Can cause esophageal injury
    b. Comes in limited sizes
    c. Cannot be used for pediatric patients
    d. All of the above

22. After an advanced airway has been placed:

    a. Keep following the 30:2 rule
    b. Interrupt chest compressions for sixty seconds
    c. deliver continuous, asynchronous ventilations once every six seconds
    d. Suction every two minutes

EXERCISE 3 – LABEL THE PICTURE

Identify and name the airway used in each of the following images:

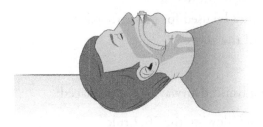

23. _____

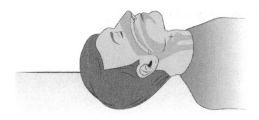

24. _____

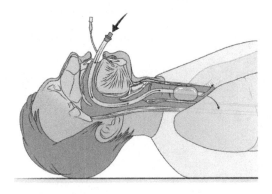

25. _____

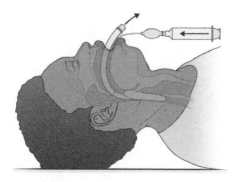

26. _____

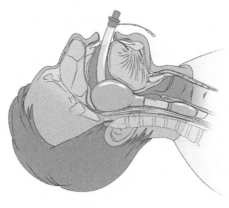

27. _____

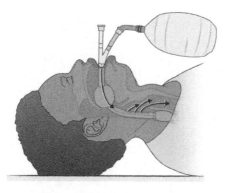

28. _____

CHAPTER 4:

ACLS Survey 4 - Defibrillation

EXERCISE 1 - FILL IN THE BLANKS

1. While CPR can sustain life by artificially circulating blood in the body, it cannot restore the original rhythm of the heart which requires _____.

2. Studies have shown that using defibrillators in conjunction with CPR can greatly improve _____.

3. For monophasic defibrillators, the AHA recommends that a shock energy of _____ be delivered during defibrillation.

4. In terms of the pattern of current flow, the AHA prefers the use of _____ defibrillators as they have greater efficiency in terminating arrhythmias.

5. It is essential that defibrillation be immediately preceded and succeeded by _____.

6. In their 2020 guidelines, the AHA recommends a _____ shock strategy rather than a series of 'stacked' shocks with progressively higher energy levels.

7. A single, sharp impact or 'punch' to the mid-sternum that may deliver a low energy shock to the heart is called a _____.

EXERCISE 2 - MCQs

Set A

Use the following four options for each of the questions or statements in this set.

      a.   Automated external defibrillator (AED)

      b.   Implantable cardioverter defibrillator (ICD)

      c.   Wearable cardioverter defibrillator (WCD)

      d.   Manual external defibrillators

8. This defibrillator is designed to be placed surgically into the patient's body.

9. These are portable and user-friendly defibrillators designed for use by lay individuals.

10. These depend on the rescuer to analyze the rhythm and deliver a shock if appropriate.

11. They are used for patients in whom the risk of cardiac arrest is high for a short period of time – such as after cardiac bypass surgery, or just following a myocardial infarction.

Set B

12. Research shows that if defibrillation is provided within 5 minutes from the onset of sudden cardiac arrest the survival rate significantly increases to:

    a. 20%
    b. 50%
    c. 80%
    d. 100%

13. After defibrillation:

    a. Resume CPR immediately
    b. Perform a post-shock rhythm analysis
    c. Switch off the defibrillator
    d. Wait for return of spontaneous circulation (ROSC)

14. To reduce transthoracic impedance, the AHA recommends that the diameter of the pads or paddles used should be at least:

    a. 2-4 cm
    b. 5-7 cm
    c. 8-12 cm
    d. 13-15 cm

15. A 65-year-old man collapsed in a bathtub full of water. He has been placed on a dry floor and CPR has begun. His chest is still wet. Defibrillation is needed. What would be the next step?

    a. Avoid defibrillation in this patient
    b. Use paddles instead of adhesive pads
    c. Quickly wipe the chest dry
    d. Use a monophasic defibrillator

CHAPTER 5:

Recognizing Rhythms that occur during Cardiac Arrest

EXERCISE 1 - MCQs

Use the following four options for all the questions or statements in this exercise.

    a.   Ventricular fibrillation (VF)
    b.   Ventricular tachycardia (VT)
    c.   Pulseless electrical activity (PEA)
    d.   Asystole

1. Which one of the above is a regular rhythm and shockable?

2. Which one of the above is an irregular rhythm and shockable?

3. In which one of the above, the ventricular rate may range from 150 to 250 beats per minute?

4. Which one of the above is also known as electromechanical dissociation?

5. Which one of the above displays as a complete flatline on the ECG monitor?

6. In which one of the above, there is discernable electrical activity but defibrillation must not be performed?

7. In which one of the above, the electrical activity is highly variable and usually reflects the underlying rhythm?

8. Identify the following rhythm:

9. Identify the following rhythm:

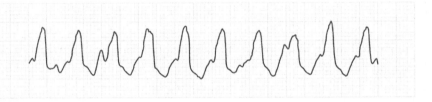

10. Identify the following rhythm:

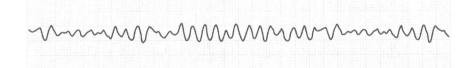

CHAPTER 6:

Other Considerations in the ACLS Algorithm

EXERCISE 1 - FILL IN THE BLANKS

Depending on the patient's medical history and the circumstances in which the cardiac arrest occurred, the most likely diagnosis must be considered and addressed.

Regarding the differential diagnosis, list the 5 Hs:

1. _____

2. _____

3. _____

4. _____

5. _____

Now, list the 5 Ts:

6. _____

7. _____

8. _____

9. _____

10. _____

EXERCISE 2 - MCQs

11. Which of the following parameters are indicative that return of spontaneous circulation (ROSC) has occurred?

    a. Visible pulse and blood pressure on the monitor
    b. Visible arterial pressure waves on the intra-arterial monitor
    c. ETCO2 values abruptly increase, usually to values above 40mmHg
    d. All of the above

12. Which one of the following is not an ALS termination of resuscitation (TOR) criterion?

    a. EMS provider did not witness the arrest
    b. The patient received bystander CPR
    c. No ROSC after complete ACLS care was provided
    d. No shock was delivered

13. For in-hospital resuscitation, in most scenarios, the patient is intubated. Therefore, ETCO2 may be used as a prognostic indicator for considering TOR. The AHA guidelines suggest that, in an intubated patient, if the ETCO2 has been less than 10 mm Hg after resuscitating for more than how many minutes would it be considered an indicator of TOR?

    a. 10
    b. 20
    c. 30
    d. 40

Post-Cardiac Arrest Care

EXERCISE 1 - FILL IN THE BLANKS

1. The moment return of spontaneous circulation (ROSC) occurs, _____ care begins.

2. While the ACLS algorithm is responsible for saving an individual's life, proper post-cardiac arrest care can ensure that the individual attains a decent _____ following cardiac arrest.

3. Cardiac arrest is a condition where the whole body essentially goes into ischemia. While ROSC restores perfusion, there is a chance for _____ injury to develop.

4. In their 2020 guidelines, the AHA recommends targeted temperature management (TTM) for all patients who survive a _____.

5. Steroids may improve survival in patients with _____ shock.

6. The process of establishing neurological outcome in a patient after cardiac arrest is called _____.

7. Several times, the cause of death following post-cardiac arrest brain injury is due to active withdrawal of life support treatment, which is based on a prognosis of predicted _____ neurological outcome.

8. In brain CT scans, reduced _____ ratio after cardiac arrest is indicative of poor neurological outcomes.

EXERCISE 2 – POST-CARDIAC ARREST CARE ALGORITHM

Complete the labeling in the post-cardiac arrest care algorithm shown below.

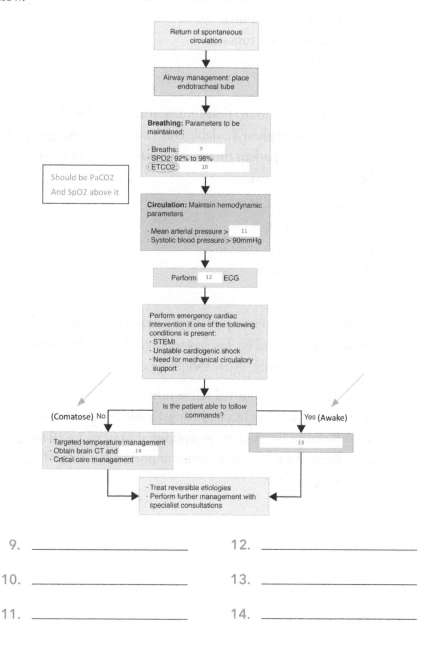

Return of spontaneous circulation

↓

Airway management: place endotracheal tube

↓

Breathing: Parameters to be maintained:
· Breaths: 9
· SPO2: 92% to 98%
· ETCO2: 10

Should be PaCO2
And SpO2 above it

↓

Circulation: Maintain hemodynamic parameters
· Mean arterial pressure > 11
· Systolic blood pressure > 90mmHg

↓

Perform 12 ECG

↓

Perform emergency cardiac intervention if one of the following conditions is present:
· STEMI
· Unstable cardiogenic shock
· Need for mechanical circulatory support

↓

Is the patient able to follow commands?

(Comatose) No / Yes (Awake)

· Targeted temperature management
· Obtain brain CT and 14
· Crtical care management

13

· Treat reversible etiologies
· Perform further management with specialist consultations

9. _____   12. _____

10. _____   13. _____

11. _____   14. _____

41

EXERCISE 3 - MCQs

15. Which one of the following is a goal of post-cardiac arrest care?

    a. Optimize ventilation and circulation
    b. Preserve heart function
    c. Preserve brain function
    d. All of the above

16. Avoiding hypoxia in the post-cardiac arrest period is a high priority. Therefore, the highest available concentration of oxygen must be delivered to the patient (preferably 100%), and SpO2 levels must be maintained between:

    a. 82% to 88%
    b. 88% to 92%
    c. 92% to 98%
    d. 98% to 100%

17. A 60-year-old man is in post-cardiac arrest care. His systolic blood pressure is 75 mmHg. He has received 2 liters of 0.9% normal saline intravenously. What should be the next step?

    a. Maintain the normal saline infusion.
    b. Give 2 liters of lactated Ringer's solution intravenously.
    c. Begin an infusion of epinephrine at a rate of 2 – 10 mcg/min.
    d. Begin an infusion of norepinephrine at a rate of 0.1 – 0.5 mcg/kg/min.

18. A post-arrest 12-lead ECG was obtained in a 66-year-old woman. A STEMI was identified. She was sent for an emergent coronary angiography which revealed a block. What procedure is this woman a candidate for?

    a. EEG
    b. PCI
    c. Defibrillation
    d. Arthroplasty

19. Which one of the following is an indication for coronary angiography in a post-cardiac arrest patient?

    a. Strong suspicion of a cardiac cause
    b. History of chest pain
    c. History of congenital heart disease
    d. All of the above

20. Targeted temperature management (TTM), also known as mild therapeutic hypothermia, is a process where hypothermia is induced in a patient who has just survived a:

    a. car accident
    b. physical trauma
    c. cardiac arrest
    d. bout of asthma

21. In TTM, the patient's body temperature is maintained at a constant core temperature between 32°C to 36°C. After the constant temperature is attained, it must be maintained for at least how many hours?

    a. 12
    b. 24
    c. 36
    d. 72

22. While TTM is advocated after a cardiac arrest, it must be performed only in the:

    a. field
    b. car
    c. ambulance
    d. hospital

23. The decision to withdraw life support in a post-cardiac arrest patient is an important one. Which one of the following is a process that helps in making that decision?

    a. Capnography
    b. Neuroprognostication
    c. Head X-ray
    d. Bereavement

24. Which one of the following is an AHA recommendation for proper neuroprognostication?

    a. It is essential in all patients who remain comatose after cardiac arrest.
    b. It should be a multimodal process and not based on any single finding.
    c. Ideally, it should be performed at least 72 hours after the patient has attained normothermia.
    d. All of the above.

25. A 72-year-old man has been comatose for more than 72 hours after cardiac arrest. Which one of the following should not be used as an indicator of poor neurological prognosis?

    a. Absence of bilateral pupillary light reflex
    b. Absence of bilateral corneal reflex
    c. Status myoclonus
    d. Absence of best motor response on the Glasgow coma scale

26. High serum values of which one of the following biomarkers is indicative of poor neurological outcome, when considered with other factors?

    a. Insulin
    b. Calcitonin
    c. Thyroid-stimulating hormone (TSH)
    d. Neuron-specific enolase (NSE)

27. Which one of the following EEG signs would indicate poor neurological prognosis in a patient who has been comatose for more than 72 hours after cardiac arrest?

    a. Bilateral absence of N-20 somatosensory evoked potential waves
    b. Persistent status epilepticus
    c. Burst suppression in the absence of sedating drugs
    d. All of the above

Identification and Management of Specific Case Scenarios

CHAPTER 1:

Identification and Management of Non-Arrest Rhythms

EXERCISE 1 - FILL IN THE BLANKS

1. _____ is a classic example of polymorphic ventricular tachycardia.

2. Healthy people, such as athletes, can have _____ bradycardia.

3. In sinus bradycardia, the rate is usually between _____ beats per minute.

4. In first-degree AV block, _____ is prolonged.

5. In second-degree AV block type I, the PR interval progressively lengthens between each cycle until the _____ disappears altogether for one cycle.

6. In third-degree AV block, atrial and ventricular rates are completely _____.

7. Tachycardia usually becomes symptomatic when the heart rate crosses _____ beats per minute.

8. In atrial fibrillation, the rhythm is _____ irregular.

9. In atrial flutter, the rhythm is _____ irregular.

10. In atrial flutter, the flutter waves show a characteristic _____ pattern.

11. In supraventricular tachycardia, the QRS complexes are _____.

12. In atrial fibrillation, if a biphasic defibrillator is used for synchronized cardioversion, the optimal energy setting is _____.

13. In atrial flutter, if a biphasic defibrillator is used for synchronized cardioversion, the optimal energy setting is _____.

14. Vagal maneuvers can be attempted to treat hemodynamically stable _____ tachycardias.

EXERCISE 2 – RHYTHM IDENTIFICATION

Identify and label the following rhythms:

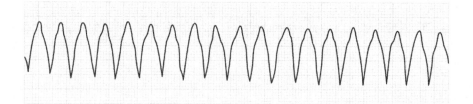

15. _____

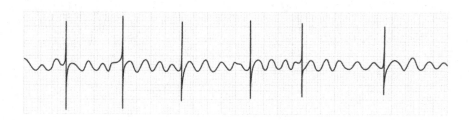

16. _____

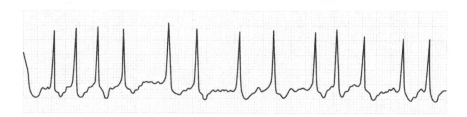

17. _____

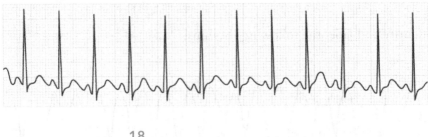

18. _____

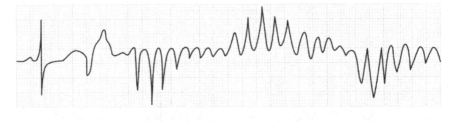

19. _____

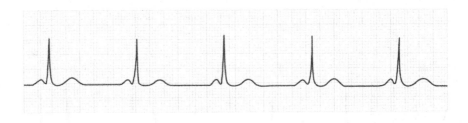

20. _____

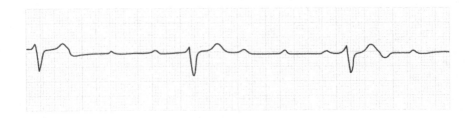

21. _____

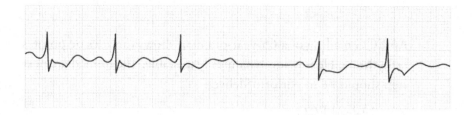

22. _____

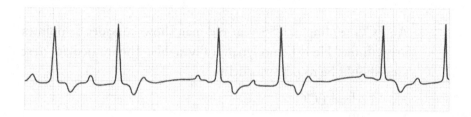

23. _____

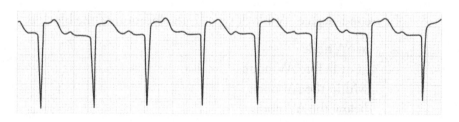

24. _____

EXERCISE 3 - MCQs

25. An ECG reading of a 65-year-old man shows torsades de pointes arrhythmia. He is hemodynamically unstable. The immediate next step should be to perform/deliver:

    a. 12-lead ECG
    b. CPR
    c. Defibrillation
    d. Amiodarone

26. An ECG reading of a 55-year-old man shows torsades de pointes arrhythmia. He is hemodynamically stable. The immediate next step should be to perform/deliver:

    a. 12-lead ECG
    b. CPR
    c. Defibrillation
    d. Amiodarone

27. In second-degree AV block type II, the location of the block is:

    a. at SA node
    b. at or before AV node
    c. within the AV node
    d. below the AV node

28. If a patient presents with symptomatic bradycardia, the cause must be ascertained and addressed. If there is hemodynamic compromise, atropine is administered to increase the heart rate. If the bradycardia does not respond to atropine, which of the following actions are appropriate:

    a. A rate accelerating agent, such as epinephrine can be administered.
    b. Alternatively, transcutaneous pacing can be considered.
    c. Patient can be prepared for emergent transvenous pacing.
    d. All of the above.

29. Which of the following conditions can result in sinus tachycardia?

    a. Fever
    b. Hyperthyroidism
    c. Anxiety
    d. All of the above

30. In patients with atrial fibrillation or atrial flutter, if they are hemodynamically unstable, the immediate next step should be to perform:

    a. CPR
    b. defibrillation
    c. cardioversion
    d. angioplasty

31. In patients with wide complex tachycardias, if they are hemodynamically stable but the cause is not immediately identifiable, which drug can be considered for IV administration?

    a. verapamil
    b. adenosine
    c. epinephrine
    d. norepinephrine

EXERCISE 4 – SYMPTOMATIC BRADYCARDIA ALGORITHM

Complete the labeling in the symptomatic bradycardia management algorithm shown below.

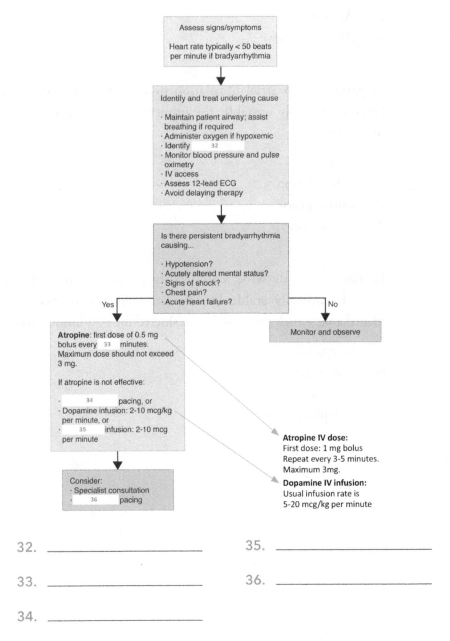

32. _____   35. _____

33. _____   36. _____

34. _____

EXERCISE 5 – SYMPTOMATIC TACHYCARDIA ALGORITHM

Complete the labeling in the symptomatic tachycardia management algorithm shown below.

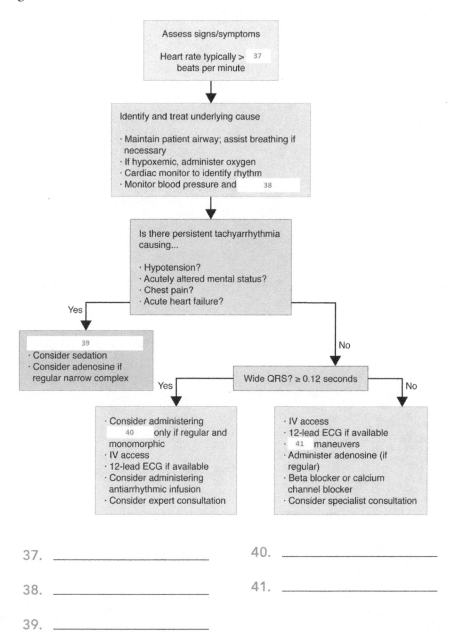

37. _____

38. _____

39. _____

40. _____

41. _____

CHAPTER 2:

Management of Opioid Overdose

EXERCISE 1 - FILL IN THE BLANKS

1. In pediatric opioid overdose, while the focus should be on providing high-quality CPR and ventilation, administration of _____ by responders is reasonable.

2. The increase in misuse of prescription as well as non-prescription opioids in the recent years has led to what is now known as the _____ epidemic.

3. _____ has been shown to significantly improve prognosis following opioid overdose. It is an opioid antagonist which works by competitively bonding to opioid receptors.

EXERCISE 2 - MCQs

4. Opioid overdoses primarily cause depression of the respiratory system and central nervous system, leading initially to:

    a. respiratory arrest
    b. cardiac arrest
    c. stroke
    d. renal failure

5. In opioid emergencies, the correct sequence of assessment and management is:

    a. C-A-B-D
    b. C-D-B-A
    c. D-C-B-A
    d. A-B-C-D

6. In a collapsed patient, which of the following scenarios should arouse suspicion of an opioid overdose?

    a. Eyewitness accounts or history of usage from bystanders, family or friends.
    b. Evidence of drug usage paraphernalia – injections or pill bottles.
    c. Patient exhibits pupillary miosis, or needle tracks on skin.
    d. All of the above.

7. A paramedical team arrives at the home of a collapsed person after receiving an emergency call. It is a 55-year-old man. His family recall a history of drug abuse. An opioid overdose is suspected. The patient shows significant improvement after naloxone administration and returns to spontaneous breathing. What should be the next step?

    a. He should be provided a sprayer with intranasal naloxone.
    b. A continuous infusion of naloxone should be started.
    c. He should be transported to a hospital for monitoring.
    d. He should be allowed to get back to his normal activities.

CHAPTER 3:

Acute Coronary Syndrome

EXERCISE 1 - FILL IN THE BLANKS

1. Acute coronary syndrome (ACS) generally encompasses all types of myocardial infarction (ST-segment elevation MI and non-ST-segment elevation MI), as well as _____.

2. The most common cause of acute coronary syndrome is a ruptured _____.

3. A patient holding a clenched fist over their chest is called _____ sign.

4. It is important to remember that patients with _____ and women may not experience the classic signs of acute coronary syndrome.

5. Emergency management of ACS must follow the acronym _____.

EXERCISE 2 – MCQs

6. The primary goals of acute coronary syndrome management are to:

    a. Reduce myocardial necrosis in order to preserve cardiac function
    b. Treat complications of acute coronary syndrome
    c. Prevent any major adverse cardiac events
    d. All of the above

7. In ACS, high flow oxygen can help alleviate hypoxic damage to the myocardium. It should be provided at the rate of:

    a. 2 L/minute
    b. 4 L/minute
    c. 6 L/minute
    d. 8 L/minute

8. In ACS, nitroglycerin must be avoided if:

    a. History of aspirin allergy
    b. Phosphodiesterase inhibitors used within the last 24 hours
    c. Systolic blood pressure is above 90 mmHg
    d. High flow oxygen is not available

9. A 12-lead ECG was obtained in a patient with suspected ACS. Which of the following are possible signs suggestive of myocardial infarction?

    a. ST-segment elevation
    b. ST-segment depression
    c. Poor progression of R waves
    d. All of the above

EXERCISE 3 – ACUTE CORONARY SYNDROME ALGORITHM

Complete the labeling in the acute coronary syndrome algorithm shown below.

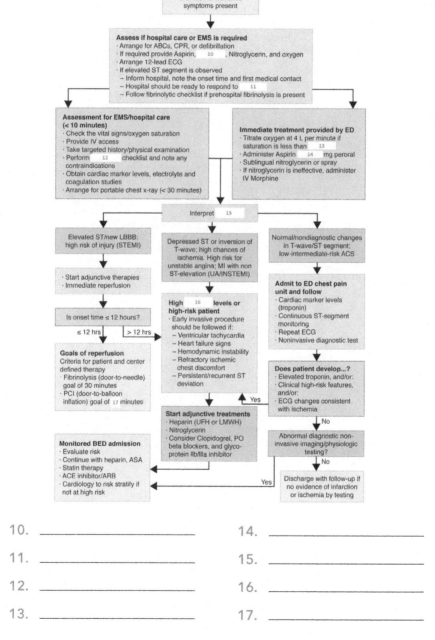

10. _____

11. _____

12. _____

13. _____

14. _____

15. _____

16. _____

17. _____

CHAPTER 4:

Acute Stroke

EXERCISE 1 - FILL IN THE BLANKS

1. Ischemic stroke is more common and results from thrombotic occlusion of the cerebral arteries, while _____ stroke occurs because of their rupture.

2. In a stroke scenario, if any family members or bystanders were present, it is important to ask them about the time of _____.

EXERCISE 2 - MCQs

3. Both ischemic stroke and hemorrhagic stroke present with similar symptoms. Which of the following symptoms can be suggestive of a stroke?

    a. Weakness in the arm, leg or face; sudden palsy
    b. Slurred speech; difficulty in speaking and forming words
    c. Blurred vision or other visual problems
    d. All of the above

4. The prognosis of stroke improves greatly if urgent medical care is received. To emphasize this, the AHA has formulated a stroke 'chain of survival'. Which one of the following is the first step in this chain?

    a. Notification of hospital and immediate transport of patient
    b. Timely management by EMS systems
    c. Activate EMS
    d. Recognize symptoms

5. Which one of the following is the second step in AHA's stroke chain of survival?

    a. Notification of hospital and immediate transport of patient
    b. Timely management by EMS systems
    c. Activate EMS
    d. Recognize symptoms

6. Which one of the following conditions can mimic a stroke and should be ruled out through a finger stick test?

    a. Hyperthyroidism
    b. Hypothermia
    c. Hypoglycemia
    d. Anemia

7. A suspected stroke patient arrives at the ED. Within how many minutes should a CT scan be performed?

    a. 10
    b. 25
    c. 45
    d. 60

8. A suspected stroke patient arrives at the ED. Within how
    many minutes should fibrinolytic therapy be initiated if the
    patient is eligible?

    a. 10
    b. 25
    c. 45
    d. 60

9. Which of the following are 'inclusion' criteria when assessing a
    stroke patient for fibrinolytic therapy?

    a. Onset of symptoms within the last three hours
    b. 18 years of age or older
    c. Acute ischemic stroke, where neurological deficit is present
    d. All of the above

10. Which one of the following is a relative exclusion criterion when
    assessing a stroke patient for fibrinolytic therapy?

    a. History of head trauma in the last three months
    b. History of stroke in the last three months
    c. Evidence of intracranial hemorrhage
    d. Surgery or trauma that occurred in the last 14 days

11. Which of the following are absolute exclusion criteria when
    assessing a stroke patient for fibrinolytic therapy?

    a. History of arterial puncture in the last 7 days
    b. Active bleeding
    c. Elevated INR
    d. All of the above

EXERCISE 3 - MATCH THE COLUMNS

The AHA has determined that delay can occur at several points along the stroke chain of survival. It refers to them as the 8 Ds. Can you match each of the 8 Ds to their correct description?

	Description	Correct match	8 Ds
12.	Rapid recognition of the symptoms of stroke		Door
13.	Immediate activation of EMS and their prompt arrival		Delivery
14.	Prompt delivery of EMS care		Disposition
15.	Deliver patient to a stroke center with proper facilities		Decision
16.	Prompt triage and evaluation of the patient, and ED management		Detection
17.	Expert evaluation and proper therapy selection		Drug
18.	Institute fibrinolytic therapy, and intra-arterial treatment		Dispatch
19.	Admit patient to critical care unit		Data

EXERCISE 4 – ACUTE STROKE ALGORITHM

Complete the labeling in the acute stroke management algorithm shown below.

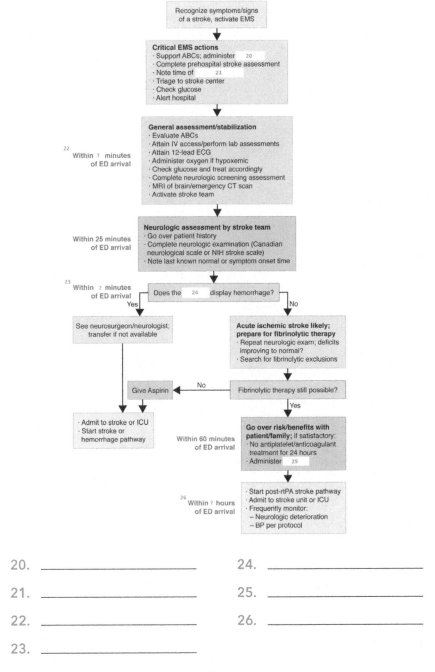

20. _____

21. _____

22. _____

23. _____

24. _____

25. _____

26. _____

ACLS in Special Populations

CHAPTER 1:

ACLS Considerations in Pediatric Patients

EXERCISE 1 - FILL IN THE BLANKS

1. If the infant/child has a pulse but inadequate respiration, provide 20 to 30 rescue breaths per minute or 1 breath every _____ seconds. The same rate is applicable when providing CPR with an advanced airway.

2. It must always be remembered that _____ are not miniature adults, a rule that applies to ACLS as well.

3. Neonates are babies that are less than _____ days old.

4. Infants are children below _____ of age.

5. Unlike adults, children usually develop _____ arrest prior to cardiac arrest.

6. For children, chest compressions must be given at a rate of _____ per minute.

7. If an advanced airway is to be placed in a child, a _____ ET tube is preferred.

8. Epinephrine must be administered within _____ minutes of beginning chest compressions.

9. If an AED is used for children below 8 years of age, a model with a _____ is preferable if available.

10. In pediatric patients, the initial dose of energy delivered during defibrillation must be _____ of body weight.

11. In pediatric patients who do not regain consciousness following ROSC, targeted _____ management is recommended.

12. During post-cardiac arrest care in pediatric patients, systolic blood pressure must be maintained above the _____ percentile for that age and gender.

13. During post-cardiac arrest care in pediatric patients, continuous EEG monitoring is needed in order to identify and manage _____.

EXERCISE 2 – PEDIATRIC ACLS ALGORITHM

Complete the labeling in the pediatric ACLS algorithm shown below.

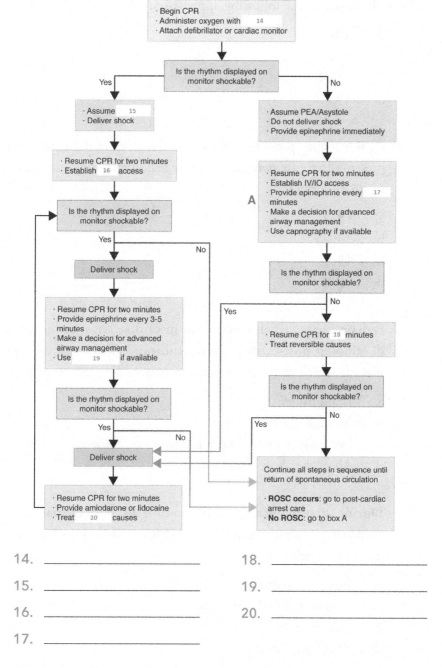

14. _____

15. _____

16. _____

17. _____

18. _____

19. _____

20. _____

EXERCISE 3 - MCQs

21. Concerning CPR in pediatric patients, the AHA recommends which one of the following approaches?

    a. A-B-C
    b. B-A-C
    c. C-A-B
    d. A-C-B

22. CPR is being provided to a child. An advanced airway is not present. There are two rescuers. What should be the compression: ventilation ratio?

    a. 15: 2
    b. 30: 2
    c. 45: 2
    d. 30: 1

23. Pediatric patients who recover from cardiac arrest are at risk of developing post-cardiac arrest syndrome. This syndrome can include which of the following components?

    a. Brain injury
    b. Myocardial dysfunction
    c. Systemic ischemia and reperfusion response
    d. All of the above

24. During neurological prognostication in post-cardiac arrest pediatric patients, which of the following EEG signs have been associated with poor outcomes?

    a. Flat EEG patterns
    b. Attenuated EEG patterns
    c. Burst suppression
    d. All of the above

CHAPTER 2:

Management of Specific Pediatric Clinical Scenarios

EXERCISE 1 - FILL IN THE BLANKS

1. In a pediatric patient presenting with shock, after each bolus of fluid administration, the patient must be reassessed both for responsiveness and signs of _____.

2. In pediatric patients with hemorrhagic shock, administration of _____ should be considered rather than crystalloids or colloids.

3. In pediatric patients with septic shock that is fluid-refractory, a vasoactive infusion of either epinephrine or _____ can be given.

4. In mild cases of foreign body airway obstruction where the child is conscious, the child should be encouraged to clear the airway by _____.

5. In cases of severe airway obstruction, the Heimlich maneuver must be attempted in children in the form of abdominal thrusts, until the object is expelled or the child becomes _____.

6. In cases of severe airway obstruction, if the child becomes unresponsive, begin _____ immediately.

7. In suspected cases of foreign body airway obstruction, under no circumstances must a _____ finger sweep be performed, as this carries the risk of pushing the object deeper into the airway.

8. In pediatric patients who develop bradycardia, if the oxygenation and ventilation is effective but bradycardia persists, _____ must begin immediately.

9. _____ refers to inflammation of the cardiac muscle, which can lead to a decrease in cardiac output and cause hypoxia to end organs.

10. In pediatric patients with single ventricle disease, due to imbalances in pulmonary and systemic blood flow and increased myocardial work, such patients are always at an increased risk of _____.

11. Pulmonary hypertensive crises in pediatric patients can lead to _____ if not handled immediately.

12. During pulmonary hypertensive crises, initial therapy should consist of prostacyclin or inhaled _____.

EXERCISE 2 - MCQs

13. Most cases of pediatric cardiac arrest, except for traumatic cardiac arrest, occur when the patient has:

    a. an underlying medical condition
    b. oxygen deprivation
    c. delayed developmental milestones
    d. premature birth

14. For pediatric patients with cardiogenic shock, early expert consultation must be obtained. If an inotropic infusion is considered, which of the following drugs can be used?

    a. Epinephrine
    b. Dopamine
    c. Dobutamine
    d. All of the above

15. A 4-year-old boy is brought to the ED with suspected respiratory failure. His breathing is inadequate but a pulse is present. Emergency intubation is about to be performed. Which drug should be given as premedication to prevent the risk of bradycardia?

    a. Atropine
    b. Epinephrine
    c. Amiodarone
    d. Cortisol

16. If the cause of bradycardia in a child is increased vagal tone, which drug should be administered?

    a. Atropine
    b. Epinephrine
    c. Amiodarone
    d. Cortisol

17. In a child with supraventricular tachycardia, if the patient is hemodynamically unstable and there is evidence of cardiovascular compromise (signs of hypotension, shock or altered mental status), which one of the following should be administered?

    a. Synchronized cardioversion
    b. Atropine
    c. Adenosine
    d. Epinephrine

18. In a child with supraventricular tachycardia, if the patient is hemodynamically stable, which one of the following should be administered?

    a. Synchronized cardioversion
    b. Atropine
    c. Adenosine
    d. Epinephrine

19. Which of the following are strategies that the AHA recommends in the management of pediatric patients with myocarditis and cardiomyopathy?

    a.   Immediate transfer to the ICU for advanced management
    b.   Extracorporeal life support
    c.   Mechanical circulatory support
    d.   All of the above

20. Single ventricle disease refers to a number of congenital conditions in which one of the ventricles is underdeveloped and may be non-functional. Patients need to undergo a series of staged surgical procedures. What is the goal of these procedures?

    a.   To provide unobstructed systemic blood circulation
    b.   To allow effective atrial communication
    c.   To optimize pulmonary blood flow so that volume load on the left ventricle is minimized
    d.   All of the above

CHAPTER 3:

ACLS Considerations in Neonates

EXERCISE 1 - FILL IN THE BLANKS

1. Prevention of _____ in the baby is the primary focus of clinical care right after birth.

2. If the newborn is apneic or has a low heart rate, _____ ventilation must be provided.

3. Positive pressure ventilation in a newborn should not last more than _____.

EXERCISE 2 - MCQs

4. The initial evaluation of a newborn baby can be performed before:
   a. cord clamping
   b. suctioning
   c. drying the baby
   d. maintaining warm temperature

5. Skin-to-skin contact between the newborn and mother can improve which of the following?

   a. Temperature control
   b. Blood glucose stability
   c. Breastfeeding ability
   d. All of the above

6. If a newborn baby is breathing spontaneously but requires respiratory support, which one of the following would be preferable?

   a. continuous positive airway pressure (CPAP)
   b. warm temperature
   c. suctioning of airways
   d. begin chest compressions

7. A newborn baby is apneic and has a heart rate of 50 beats per minute. Positive pressure ventilation is provided but the heart rate remains below 60. What should be the next step?

   a. continuous positive airway pressure (CPAP)
   b. warm temperature
   c. suctioning of airways
   d. begin chest compressions

8. In newborns, which one of the following is the best option for intravenous access?

   a. femoral vein
   b. subclavian vein
   c. umbilical vein
   d. inferior vena cava

CHAPTER 4:

ACLS Considerations in Pregnant Patients

EXERCISE 1 - FILL IN THE BLANKS

1. If a woman suffers a cardiac arrest during pregnancy, focus must remain on the resuscitation of the _____.

2. When commencing CPR in a pregnant patient, manual displacement of the _____ must be performed to the left side before delivering chest compressions.

3. Studies have shown that perimortem cesarean delivery (PMCD) can improve fetal outcomes if undertaken when the fetus is greater than _____ weeks of age.

EXERCISE 2 - MCQs

4. In pregnant patients who suffer a cardiac arrest, oxygen and airway management must be prioritized as these patients are more prone to:

   a. hypothermia
   b. hypoglycemia
   c. hypoxia
   d. hypokalemia

5.   Which of the following could cause maternal cardiac arrest during childbirth?

    a.   Hemorrhage

    b.   Amniotic fluid embolism

    c.   Sepsis

    d.   All of the above

6.   A pregnant woman needs CPR. An advanced airway is in place. While performing continuous chest compressions, one rescue breath should be delivered every how many seconds?

    a.   3

    b.   6

    c.   9

    d.   12

7.   Maternal resuscitation is in progress for a pregnant patient. Should you attempt fetal monitoring?

    a.   Yes, to keep a check on the vitals of the fetus

    b.   No, as it can interfere with maternal resuscitation

    c.   Continue fetal monitoring if it was already set up

    d.   Consult with an obstetrician

8.   As with other patients, targeted temperature management (TTM) can be undertaken for pregnant patients who remain comatose after ROSC. However, during TTM, there must be continuous fetal monitoring. This is to keep an eye out for fetal:

    a.   bradycardia

    b.   hypoxia

    c.   movements

    d.   position

CHAPTER 5:

ACLS Considerations in Patients with Known or Suspected COVID-19 Infection

EXERCISE 1 - FILL IN THE BLANKS

1. Of those who contract COVID-19 infection, 3 to 6% become critically ill and are at risk of developing hypoxemic respiratory failure, which may lead to _____.

2. For ACLS in patients with known or suspected COVID-19 infection, the AHA recommends limiting provider exposure to COVID-19 by using _____ and mechanical CPR devices.

3. For ACLS in patients with known or suspected COVID-19 infection, the AHA recommends prioritization of oxygenation and ventilation, while reducing _____.

EXERCISE 2 - COVID-19 ACLS ALGORITHM

Complete the labeling in the COVID-19 ACLS algorithm shown below.

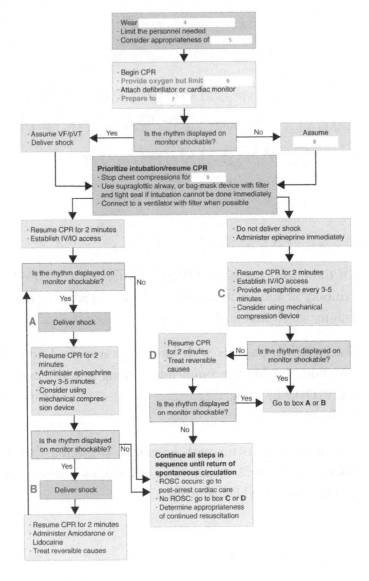

4. _____   7. _____

5. _____   8. _____

6. _____   9. _____

Answers

UNIT I
The ACLS Chain of Survival

EXERCISE 1 - FILL IN THE BLANKS

1. survival
2. recognition
3. alert/activate
4. prevent
5. rapid response
6. code blue
7. CPR
8. defibrillator
9. automated external defibrillator (AED)
10. manual
11. preserve
12. recovery

EXERCISE 2 - MCQs

13. b
14. c
15. a
16. c
17. c
18. c
19. b
20. b
21. a
22. a
23. a

UNIT II
Anatomy and Physiology of the Heart and Normal Electrocardiogram

EXERCISE 1 - FILL IN THE BLANKS

1. deoxygenated
2. left
3. cardiac cycle
4. aorta
5. P
6. T
7. R
8. 0.12 to 0.2
9. 0.04 to 0.1
10. ectopic

EXERCISE 2 - MCQs

11. c
12. c
13. c
14. c
15. b
16. d
17. b
18. a

EXERCISE 3 - LABEL THE PICTURE

19. Left atrium
20. Left bundle branch
21. Left ventricle
22. Right ventricle
23. Right bundle branch
24. Bundle of His
25. Atrioventricular (AV) node
26. Right atrium
27. Sinoatrial (SA) node

UNIT III
THE ACLS SURVEY
Chapter 1: The Adult ACLS Algorithm

EXERCISE 1 - FILL IN THE BLANKS

Set A

1. Chest compressions
2. Gain parenteral access to deliver ACLS medication
3. Basic airway management
4. Assess airway, perform advanced airway management
5. Rescue breathing
6. Check for breath sounds; Deliver 100% oxygen; Capnography to assess CPR quality
7. Use of AED if available at the scene
8. Assess heart rhythm, use defibrillator as indicated; Assess and treat reversible causes

Set B

9. CPR
10. circulation

EXERCISE 2 - ADULT ACLS ALGORITHM

11. oxygen
12. PEA/Asystole
13. epinephrine
14. two
15. IV/IO
16. capnography
17. reversible
18. shock
19. amiodarone
20. lidocaine

Chapter 2: ACLS Survey 1 - Circulation

EXERCISE 1 - FILL IN THE BLANKS

Set A

1. intraosseous
2. vasopressor
3. recoil
4. 30:2
5. interrupted
6. CPR
7. intrathoracic

Set B

8. proximal
9. three
10. bolus
11. α-adrenergic

12. epinephrine
13. vasopressin
14. lidocaine

EXERCISE 2 - MCQs

Set A

15. b
16. c
17. b
18. b
19. b
20. d
21. c

Set B

22. c
23. d
24. a
25. d
26. d
27. d
28. d

Chapter 3: ACLS Survey 2 & 3 - Airway, Breathing and Ventilation

EXERCISE 1 - FILL IN THE BLANKS

1. jaw-thrust
2. 100%
3. unresponsive
4. nasopharyngeal
5. 500 to 600 ml
6. breath

7. laryngeal mask airway
8. esophageal-tracheal
9. waveform capnography

EXERCISE 2 - MCQs

10. c
11. d
12. a
13. d
14. b
15. c
16. d
17. b
18. b
19. c
20. b
21. d
22. c

EXERCISE 3 - LABEL THE PICTURE

23. Oropharyngeal airway
24. Nasopharyngeal airway
25. Endotracheal tube
26. Laryngeal mask airway
27. Laryngeal tube
28. Combitube

Chapter 4: ACLS Survey 4 - Defibrillation

EXERCISE 1 - FILL IN THE BLANKS

1. defibrillation
2. survival/survival rates
3. 360 J

4. biphasic
5. CPR
6. single
7. precordial thump

EXERCISE 2 - MCQs
Set A

8. b
9. a
10. d
11. c

Set B

12. b
13. a
14. c
15. c

Chapter 5: Recognizing Rhythms that occur during Cardiac Arrest

EXERCISE 1 - MCQs

1. b
2. a
3. b
4. c
5. d
6. c
7. c
8. d
9. b
10. a

Chapter 6: Other Considerations in the ACLS Algorithm

EXERCISE 1 - FILL IN THE BLANKS

1. Hypovolemia
2. Hypoxia
3. Hydrogen ion acidosis
4. Hypothermia
5. Hypo/hyperkalemia
6. Tension pneumothorax
7. Tamponade, cardiac
8. Thrombosis, pulmonary
9. Thrombosis, cardiac
10. Toxins

EXERCISE 2 - MCQs

11. d
12. b
13. b

UNIT IV POST-CARDIAC ARREST CARE

EXERCISE 1 - FILL IN THE BLANKS

1. post cardiac-arrest
2. quality of life
3. reperfusion
4. cardiac arrest
5. septic
6. neuroprognostication
7. poor/negative
8. grey:white

EXERCISE 2 - POST-CARDIAC ARREST CARE ALGORITHM

9. 10 per minute
10. 35 to 45 mmHg
11. 65 mmHg
12. 12 lead
13. Critical care management
14. EEG

EXERCISE 3 - MCQs

15. d
16. c
17. c
18. b
19. d
20. c
21. b
22. d
23. b
24. d
25. d
26. d
27. d

93

UNIT V
IDENTIFICATION AND MANAGEMENT OF SPECIFIC CASE SCENARIOS

Chapter 1: Identification and Management of Non-Arrest Rhythms

EXERCISE 1 - FILL IN THE BLANKS

1. Torsades de pointes
2. physiological
3. 40 to 60
4. PR interval
5. QRS complex
6. dissociated
7. 150
8. irregularly
9. regularly
10. sawtooth
11. normal/narrow
12. 120 to 200 J
13. 50 to 100 J
14. supraventricular/narrow-complex

EXERCISE 2 - RHYTHM IDENTIFICATION

15. Monomorphic ventricular tachycardia
16. Atrial flutter
17. Atrial fibrillation
18. Sinus tachycardia
19. Torsades de pointes
20. Sinus bradycardia
21. Third degree AV block
22. Second degree AV block Type II
23. Second degree AV block - Type I
24. First degree AV block

EXERCISE 3 - MCQs

25. c
26. a
27. d
28. d
29. d
30. c
31. b

EXERCISE 4 - SYMPTOMATIC BRADYCARDIA ALGORITHM

32. cardiac rhythm
33. 3 to 5
34. Transcutaneous
35. Epinephrine
36. Transvenous

EXERCISE 5 - SYMPTOMATIC TACHYCARDIA ALGORITHM

37. 150
38. pulse oximetry
39. Synchronized cardioversion
40. adenosine
41. Vagal

Chapter 2: Management of Opioid Overdose

EXERCISE 1 - FILL IN THE BLANKS

1. naloxone
2. opioid
3. Naloxone

EXERCISE 2 - MCQs

4. a
5. d
6. d
7. c

Chapter 3: Acute Coronary Syndrome

EXERCISE 1 - FILL IN THE BLANKS

1. unstable angina
2. atherosclerotic plaque
3. Levine's
4. diabetes
5. MONA: Morphine, oxygen, nitroglycerin, and aspirin

EXERCISE 2 - MCQs

6. d
7. b
8. b
9. d

EXERCISE 3 - ACUTE CORONARY SYNDROME ALGORITHM

10. morphine
11. STEMI
12. fibrinolytic
13. 94%
14. 160-325
15. ECG
16. troponin
17. 90

Chapter 4: Acute Stroke

EXERCISE 1 - FILL IN THE BLANKS

1. hemorrhagic
2. onset of symptoms/ symptom onset

EXERCISE 2 - MCQs

3. d
4. d
5. c
6. c
7. b
8. d
9. d
10. d
11. d

EXERCISE 3 – MATCH THE COLUMNS

12. Detection
13. Dispatch
14. Delivery
15. Door
16. Data
17. Decision
18. Drug
19. Disposition

EXERCISE 4 – ACUTE STROKE ALGORITHM

20. oxygen
21. symptom onset
22. 10
23. 45
24. CT scan
25. rtPA
26. 3

UNIT VI
ACLS IN SPECIAL POPULATIONS

Chapter 1: ACLS Considerations in Pediatric Patients

EXERCISE 1 – FILL IN THE BLANKS

1. 2 to 3
2. children
3. 30
4. 1 year
5. respiratory
6. 100 to 120
7. cuffed
8. 5
9. pediatric attenuator
10. 2 J/kg
11. temperature
12. fifth
13. seizures

EXERCISE 2 – PEDIATRIC ACLS ALGORITHM

14. bag-mask
15. VF/pVT
16. IV/IO
17. 3-5
18. two
19. capnography
20. reversible

EXERCISE 3 – MCQs

21. c
22. a
23. d
24. d

Chapter 2: Management of Specific Pediatric Clinical Scenarios

EXERCISE 1 – FILL IN THE BLANKS

1. fluid overload
2. blood products
3. norepinephrine
4. coughing
5. unresponsive

6. CPR
7. blind
8. CPR
9. Myocarditis
10. cardiac arrest
11. cardiac arrest
12. nitric oxide

EXERCISE 2 - MCQs

13. a
14. d
15. a
16. a
17. a
18. c
19. d
20. d

Chapter 3: ACLS Considerations in Neonates

EXERCISE 1 - FILL IN THE BLANKS

1. hypothermia
2. positive pressure
3. 30 seconds

EXERCISE 2 - MCQs

4. a
5. d
6. a
7. d
8. c

Chapter 4: ACLS Considerations in Pregnant Patients

EXERCISE 1 - FILL IN THE BLANKS

1. mother
2. uterus
3. 20

EXERCISE 2 - MCQs

4. c
5. d
6. b
7. b
8. a

Chapter 5: ACLS Considerations in Patients with Known or Suspected COVID-19 Infection

EXERCISE 1 - FILL IN THE BLANKS

1. cardiac arrest
2. personal protective equipment (PPE)
3. aerosolization

EXERCISE 2 - COVID-19 ACLS ALGORITHM

4. personal protective equipment
5. resuscitation
6. aerosolization
7. intubate
8. PEA/asystole
9. intubation

Made in the USA
Monee, IL
12 July 2024